SIMMS LIBRARY
Albuquerque Academy
6400 Wyoming Blvd. N.E.
Albuquerque, NM 87109

Sea sOup

ZOOPLANKTON

MARY M. CERULLO

PHOTOGRAPHY
BY BILL CURTSINGER

Tilbury House, Publishers • Gardiner, Maine

Gulf of Maine Aquarium • Portland, Maine

The moment

they dropped

into the dark,

alien world,

the searchers knew

they were not alone.

Down they went, deeper and deeper into the darkness. They knew sea creatures were all around them.

When the divers finally switched on their underwater lights, they were quickly surrounded by swarms of tiny animals drawn to their lamps like insects to a porch light. The divers found what they had come for—or rather, it had found them! All around them danced the creatures that nourish most other animals in the sea—zooplankton (ZOE-o-plank-ton).

Here in the beams of light were thousands of tiny creatures, some called copepods (KO-poh-pods), related to crabs and lobsters, and others that were the baby crabs and lobsters themselves. The baby forms of other sea animals, along with creatures with odd names like krill, salps, and sea wasps, are all zooplankton—small drifting animals. (Other zooplankton, such as rotifers, water fleas, baby insects, and some copepods live in fresh water.)

592.1776
CER
2001

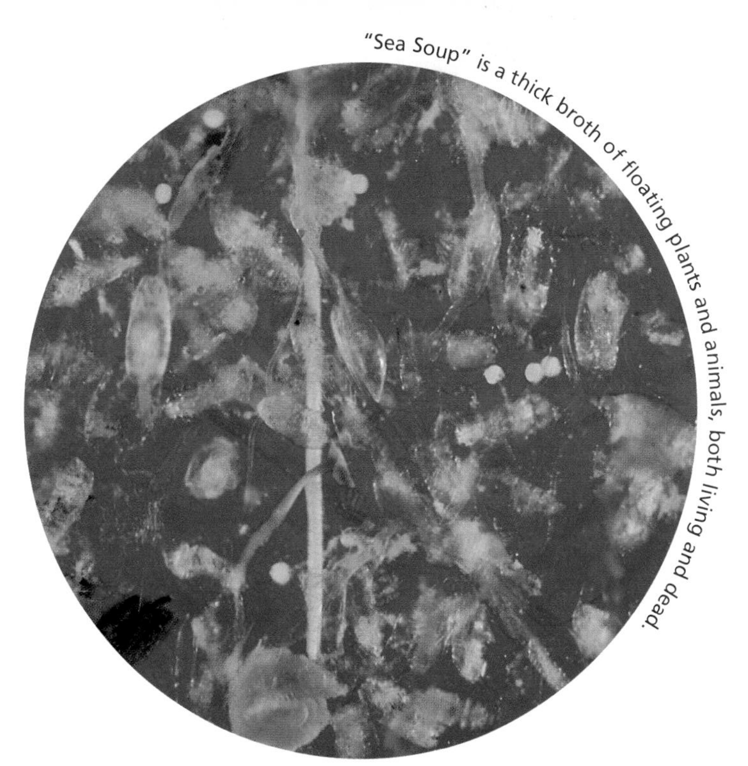

"Sea Soup" is a thick broth of floating plants and animals, both living and dead.

Plankton are named from the Greek word for "wanderer" because they drift at the mercy of the waves, tides, and currents. There are zooplankton (drifting animals) and phytoplankton (drifting plants). One of either is called a plankter, but several are plankters or plankton. Phytoplankton (FIE-toe-plank-ton) are sometimes called "sea soup" because they feed, directly or indirectly, most of the animals of the sea. Phytoplankton get their energy from the sun; animals, including the tiniest zooplankton, get their energy from capturing and eating food. Zooplankton are also essential ingredients in the ocean food chain, and we can think of them as the "chunky parts" of sea soup, because they are usually quite a bit bigger than phytoplankton.

How Does Your Ocean Garden Grow?

Most zooplankton feed on phytoplankton, but how do phytoplankton get their food? They make it themselves, as all plants do. Phytoplankton and other plants have a chemical called chlorophyll that captures the sunlight and changes it into food—sugars and other carbohydrates. Combine sunlight, water, and carbon dioxide, plus nutrients like phosphorus and nitrogen (the same fertilizers we put on our gardens), and you soon have huge "blooms" of phytoplankton. This chemical reaction, called photosynthesis, also makes oxygen that all animals need to breathe.

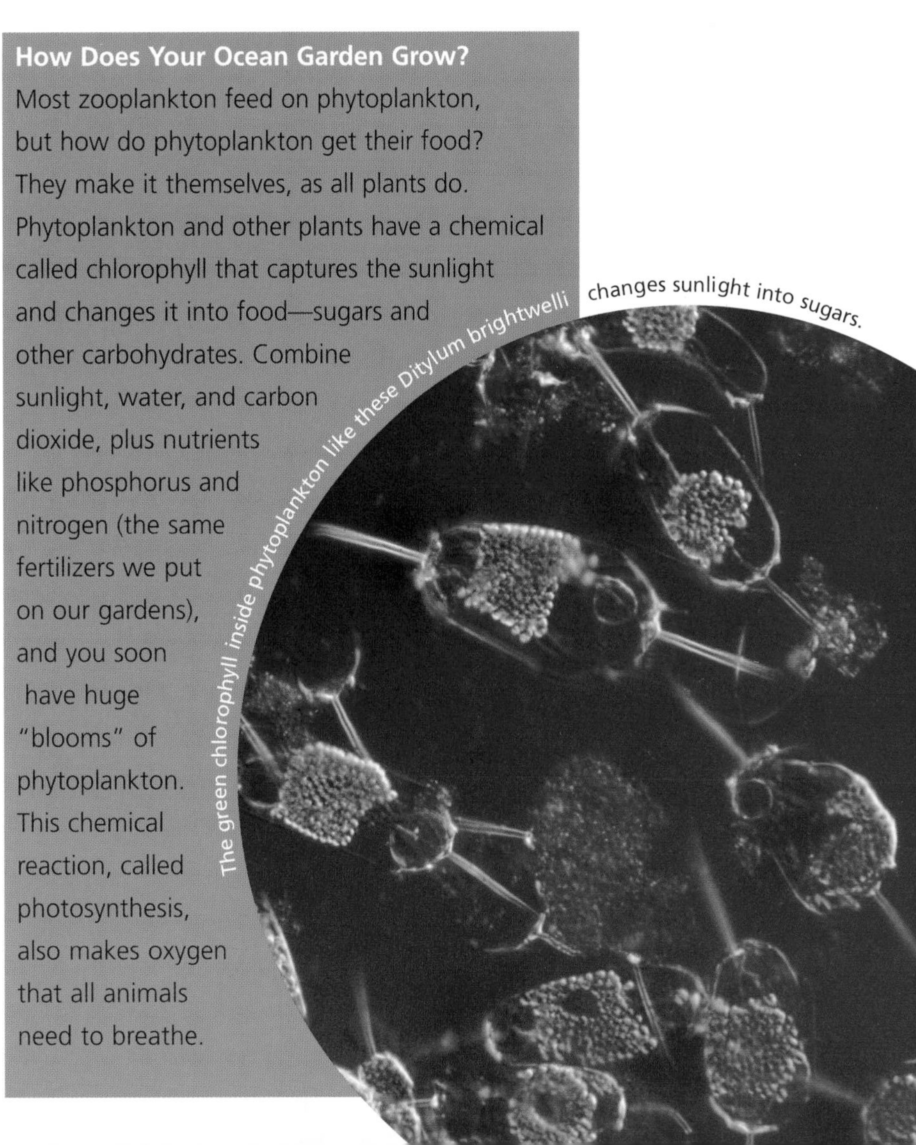

The green chlorophyll inside phytoplankton like these Ditylum brightwelli changes sunlight into sugars.

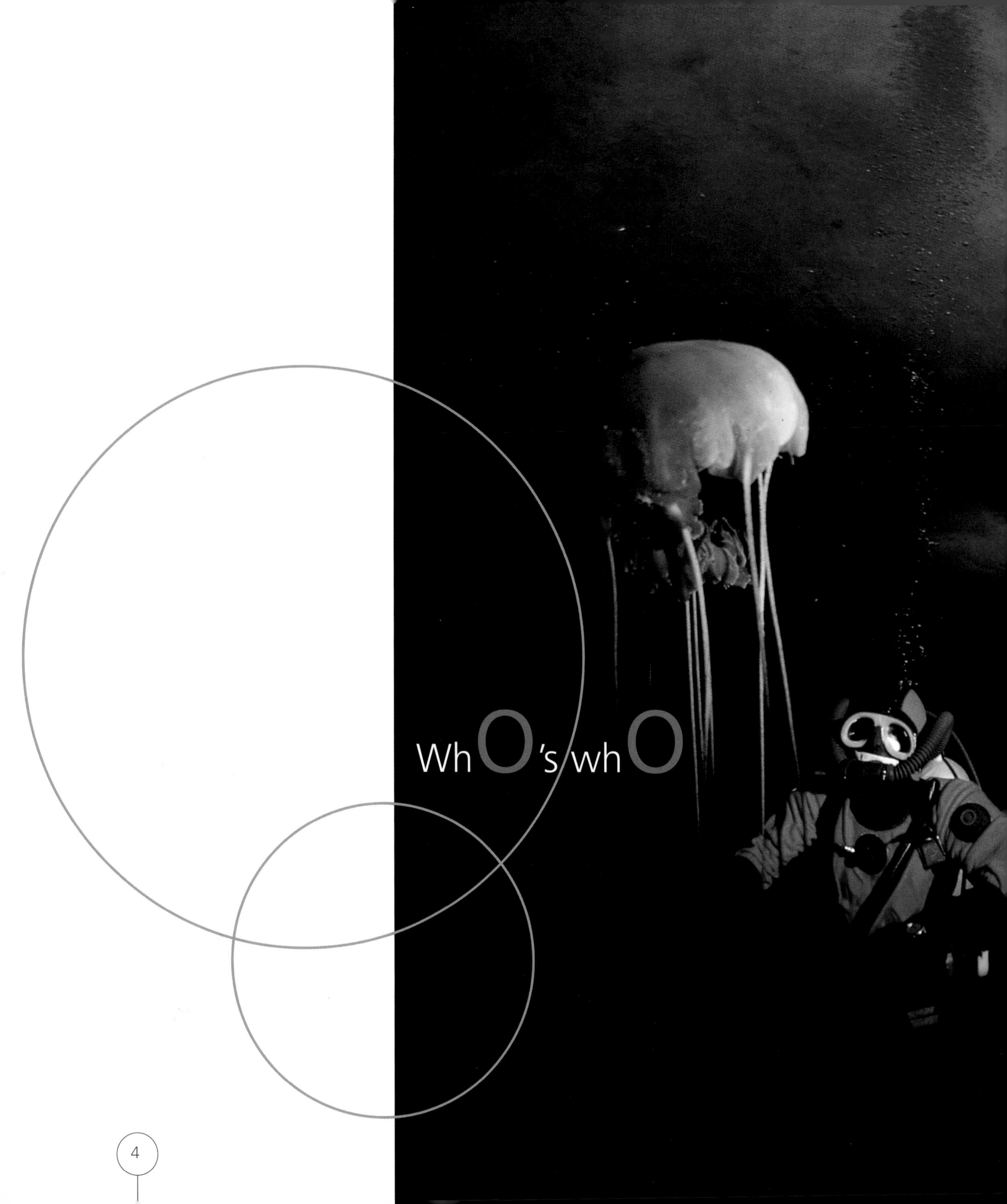

WhO's whO

in a zooplankton ZOO?

Size has nothing to do with being a zooplankter. Zooplankters can be tiny copepods or giant jellyfish.

Shaped like a spaceship, this helmet jellyfish floats in the darkness of deep (inner) space.

Any aquatic animals that are carried by

That includes the eggs of sea animals, their newly hatched offspring, many small creatures, and even some big ones.

Most phytoplankton are smaller than the period at the end of this sentence, so small that they can only be seen through a microscope. Most zooplankton can be seen with an ordinary magnifying lens. Many are big enough that we can see them zipping around a jar of sea water like turbo-charged dust particles. Unlike phytoplankton, which have to stay near the surface where the sun's light can reach them, zooplankton can live from the surface of the sea to the bottom of the deepest ocean.

Even though they are swollen with eggs, these copepods can slip through the eye of a needle.

the currents **can be called zooplankton.**

Wherever zooplankton live in the ocean, they have to find ways to float easily. Some have long spines or feathery hairs that let them spread their body weight over a bigger area. Others, like the Portuguese man-of-war, have air-filled floats, while others contain oil that makes them more buoyant. One open-ocean snail clings to a self-made raft of air bubbles. Some kinds of zoo-plankters link themselves together in long chains, making them look like skydivers holding hands as they float through the blue.

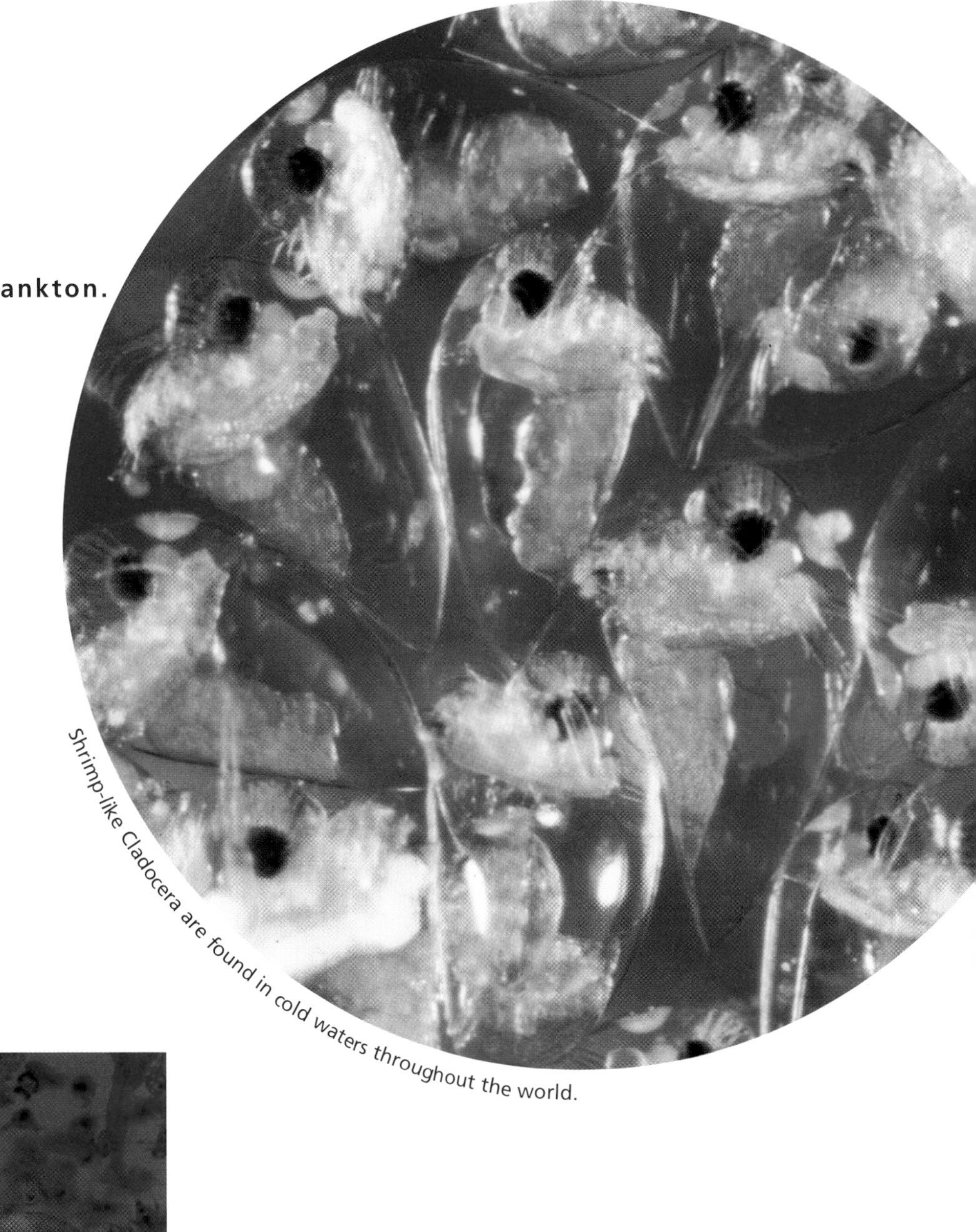

Shrimp-like Cladocera are found in cold waters throughout the world.

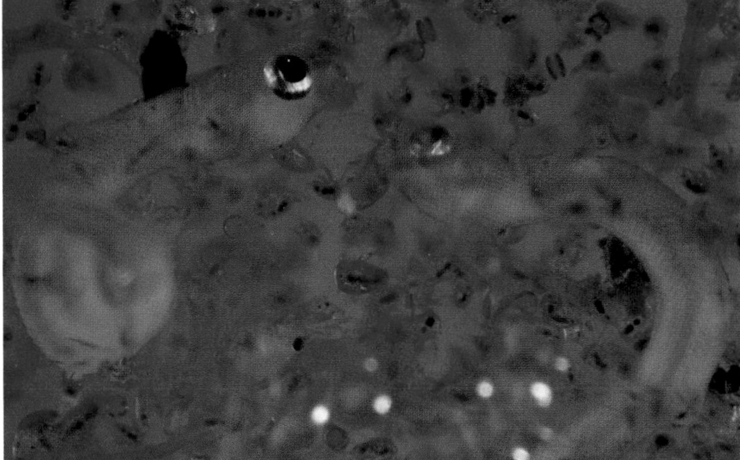

These herring larvae will be called sardines when they get big enough to fit in a can.

Who are the

Ugly Ducklings of the ocean?

Baby shrimp are almost transparent when they are zooplankton. An adult sand shrimp relies on camouflage to hide it from predators on the ocean bottom.

A fish larva looks all eyes!

Except for marine mammals, almost all th

The Ugly Duckling is the story of a hatchling who doesn't resemble its mother at all. Because it is different, it is rejected by neighbors, siblings, and strangers until it finally grows up to become a beautiful swan. In the sea, it's very common for newly hatched offspring not to look like their parents. It's just a stage—in fact, it's called the larval stage.

Many marine animals, such as fish, crabs, lobsters, barnacles, snails, clams, sea stars, and squid, go through several shape changes before they grow

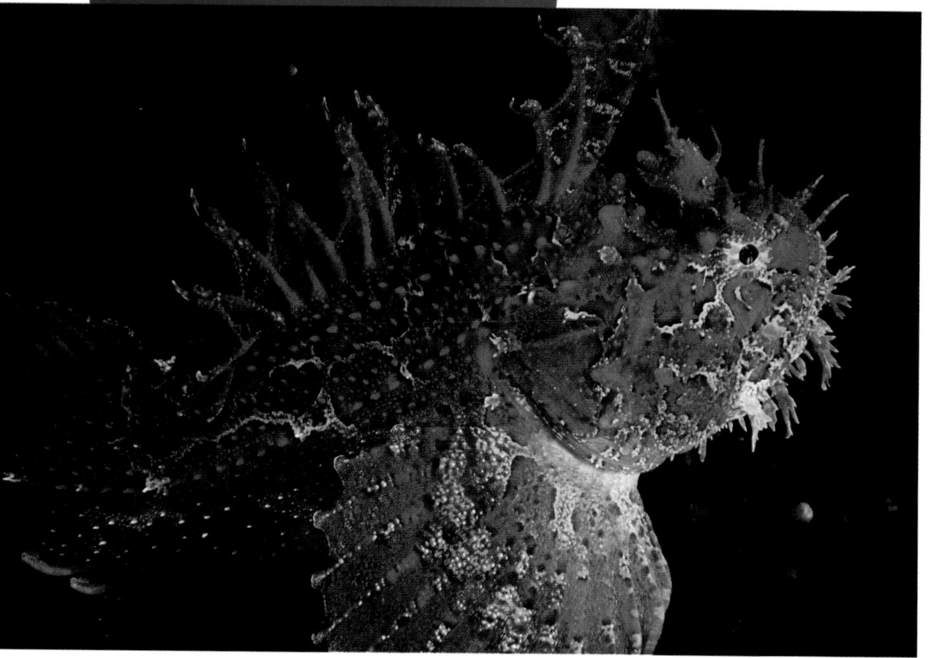

Just by looking at it, you might be able to guess that this sea raven spends most of its time hiding in seaweed on the ocean floor.

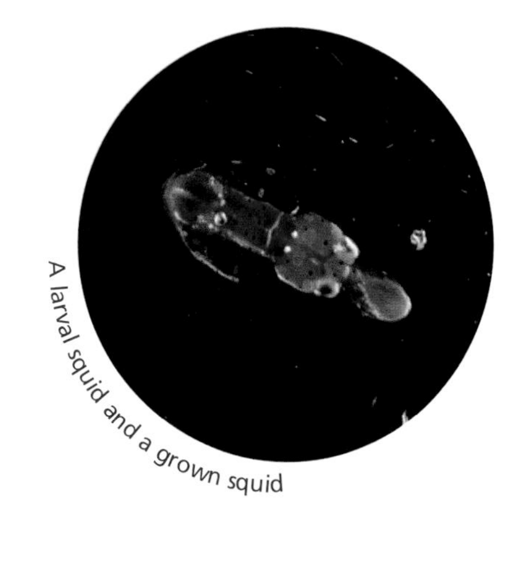

A larval squid and a grown squid

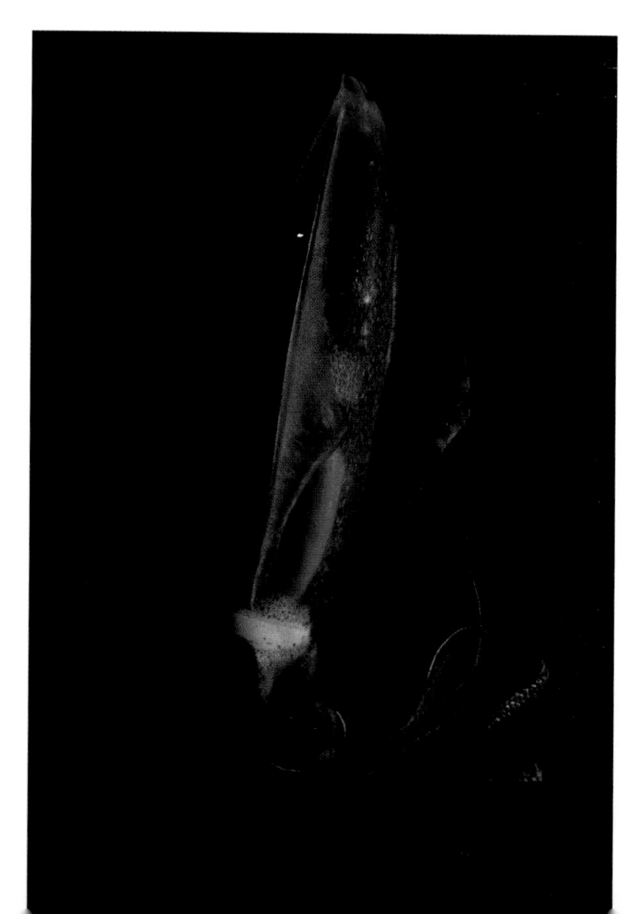

nimals in the sea start off as "Ugly Ducklings."

A baby crab is called a zoea. It is also known as the "unicorn of the sea" because of the sharp spike on its head.

Below, an adult rock crab.

into their adult forms. Their larval stages may last a few hours or several months, so these zooplankton are called temporary plankton. During that time they stay near the ocean's surface, feeding on phytoplankton or being carried by the currents to new homes. Eventually they either sink to the ocean floor and become bottom-hugging crabs, lobsters, worms, clams, or sea stars, or they may turn into strong swimmers like cod and herring. In the ocean, it's okay to be an Ugly Duckling.

There are also zooplankton that never stop being plankton. These are permanent plankton and they spend their entire lives drifting with the currents. While temporary plankton tend to live near the coast, permanent plankton often live in the open ocean. That may be one reason most of us are less familiar with permanent zooplankton like copepods, krill, sea butterflies, and salps.

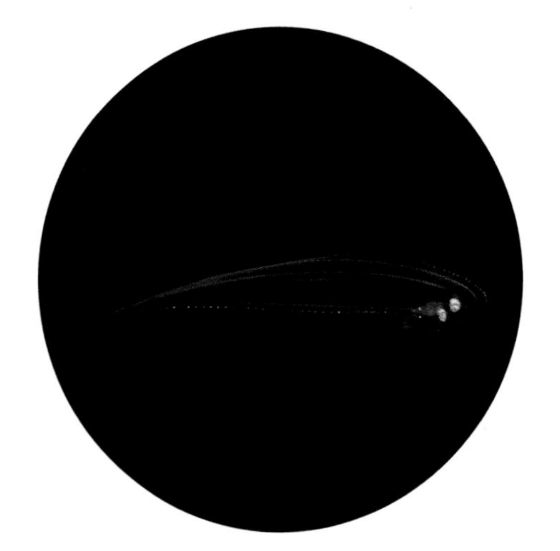

A planktonic flounder looks like any other fish, but by the time a flounder settles into life on the ocean floor, both eyes have moved to one side of its head, so it can lie flat on the sand and still see with both eyes.

Who eats

Adelie penguins and a crab-eater seal eat the same food—**krill.**

whom?

Do you eat **zooplankton?**

The word "krill" is Norwegian for "whale food."

You probably won't find zooplankton on the menu of your favorite restaurant (although some Chinese restaurants do serve jellyfish). Of course, if you eat crabs, lobsters, shrimp, or fish, you could claim you've eaten zooplankton—you just gave them time to grow up!

But some shrimp-like zooplankton called krill are the favorite food of Antarctic penguins. Krill are so abundant in the waters of the Antarctic ocean that they sometimes turn the water pink like their shells. Each krill is only the size of your thumb or smaller, yet masses of them feed almost all the animals in the Antarctic food chain. Krill are eaten by penguins, seabirds, fishes, squid, seals, and by some of the largest animals in the world—humpbacks and blue whales. A single blue whale may devour up to eight tons of krill a day.

And sometimes krill are eaten by adventurous humans. In the early 1900s, German scientists on a polar expedition to Antarctica caught and ate krill. One wrote in his journal that these two-inch-long shrimp "tasted quite good, but were rather small and tiresome to peel." In the 1970s fishermen from the Soviet Union, Japan, Poland, East Germany, South Korea, and Taiwan began harvesting krill. Today a few countries still catch krill to use for fertilizer or for fish food in sea farming operations, but humans never really developed a taste for krill-cheese or Siberian krill dumplings.

This whale shark (a fish, not a mammal) feeds exclusively on phytoplankton and zooplankton.

Krill are important to the Antarctic food chain.

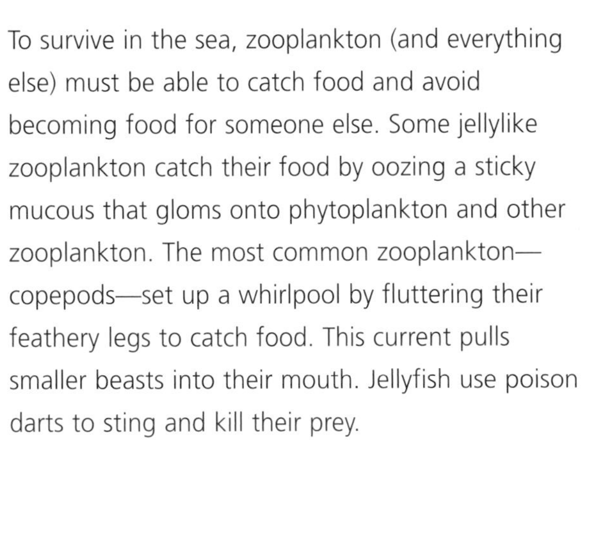

To survive in the sea, zooplankton (and everything else) must be able to catch food and avoid becoming food for someone else. Some jellylike zooplankton catch their food by oozing a sticky mucous that gloms onto phytoplankton and other zooplankton. The most common zooplankton—copepods—set up a whirlpool by fluttering their feathery legs to catch food. This current pulls smaller beasts into their mouth. Jellyfish use poison darts to sting and kill their prey.

How do zooplankton keep from being eaten? Baby crabs (called zoea) have spines that may make them too big a mouthful for other zooplankton to swallow. Baby snails (called veligers) have "wings" that sometimes help them move out of the way of predators, and copepods flick their long antennae to dart away from danger. Baby crabs and lobsters are nearly as clear as glass, so they are almost invisible to hungry fishes. Many zooplankton hide from their enemies by sinking into deeper, darker water. Jellyfish use their stinging cells to try to discourage even the most determined predators, including humans. Although zooplankton have many different ways to keep from being eaten, most zooplankton do not escape their fate—to become food for the other animals in the sea.

A lion's mane jellyfish snares a fish with its tentacles.

What is the *fastest* animal in the world?

Shaped like a bullet with antennae,

the copepod, **a tiny cousin of lobsters and crabs,**

is for its size the fastest animal on earth.

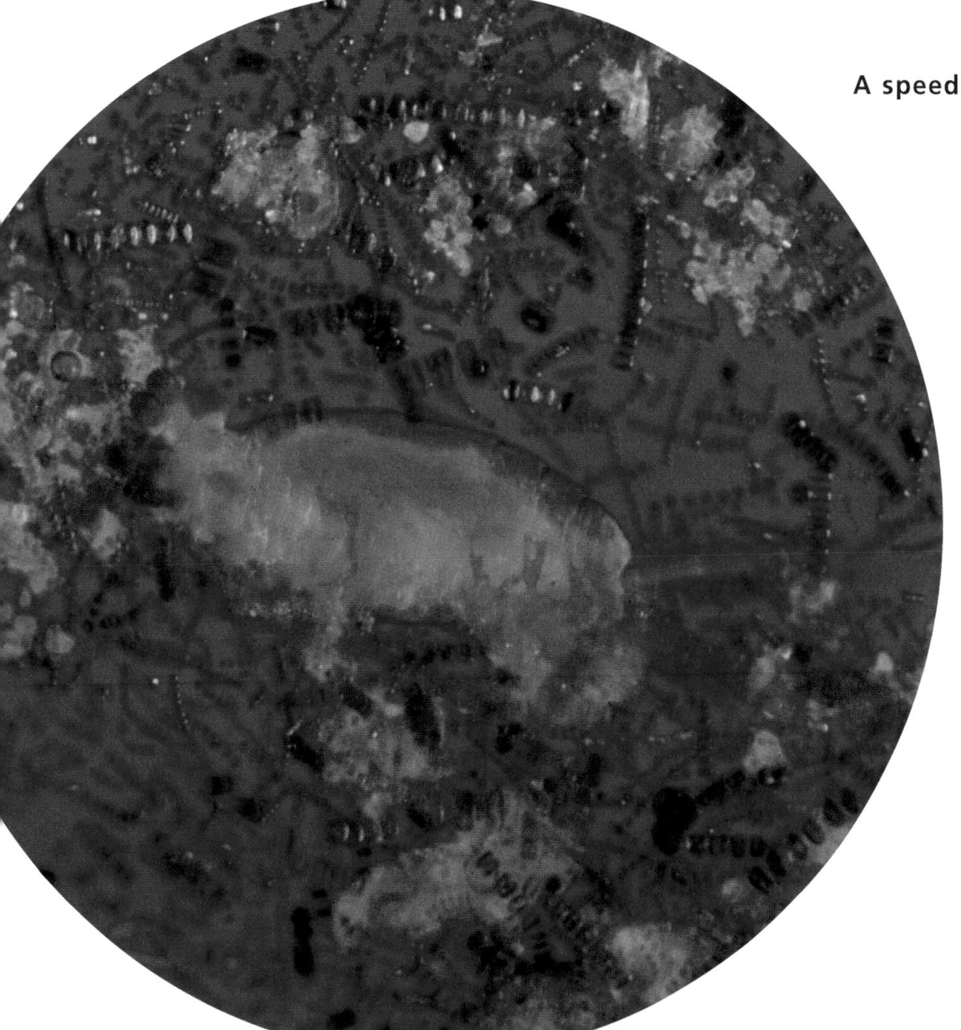

A speeding copepod looks like it's being *shot* out of a cannon!

If a cheetah (the fastest animal on land) and a copepod were the same size, a cheetah running at 70 miles per hour (mph) would compare to a copepod moving at 2,000 mph. The most distinctive feature of the copepod is its long, flowing antennae which help the animal sense the vibrations from approaching food or enemies. A copepod in search of a meal stretches out its antennae like radar, waiting for signals from the sea. If it senses danger, it slaps its antennae against its body and darts away. It can propel itself up to 500 times its body length in less than a second.

Copepods, besides being fast, are numerous, so common that they are called the "insects of the sea." Chances are they are the most abundant animals in the ocean, possibly on earth, with an estimated population of one quintillion (count the

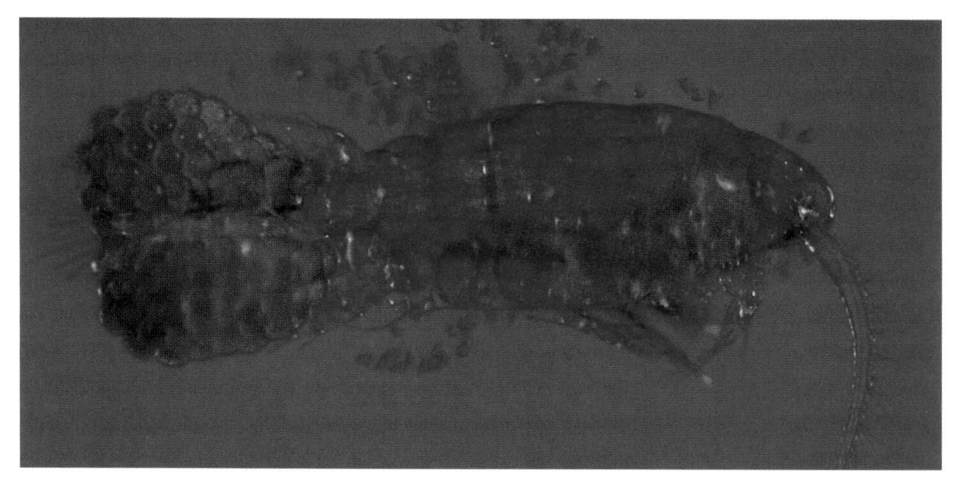

This copepod tows two large sacs of eggs behind it for about a week until the eggs hatch and scatter into the sea.

zeros—1,000,000,000,000,000,000!). If all the copepods in the world were divided equally among the entire human population, there would be enough for every person on earth to have one billion copepods. What would we do with them all? We would feed almost all the creatures in the ocean, of course. Copepods are key animals in the ocean food web. They are the main food for many larger animals in the sea, including shrimp, baby cod and haddock, seabirds, and even basking sharks and humpback whales. Over 60,000 copepods were found inside the stomach of a herring, and a right whale can eat millions of copepods every day.

The "whale lice" on this Southern right whale were plankton before they attached themselves to the whale.

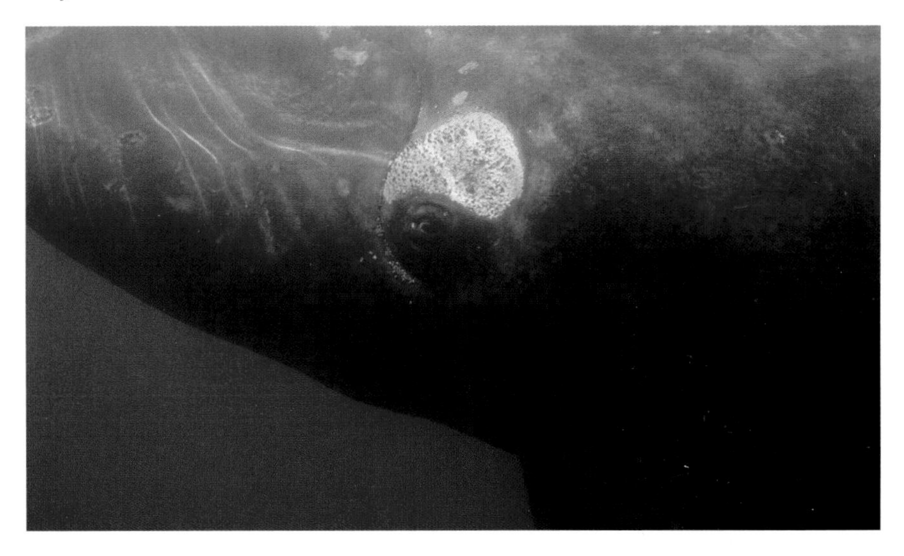

There are at least 12,000 different kinds of copepods swimming through the ocean and living in the deepest ocean trenches. Some burrow into the mud. Most are about the size of a grain of rice, but some inch-long isopods, called "whale lice,"even make their homes on whales, and others live in the mouths and gills of fish.

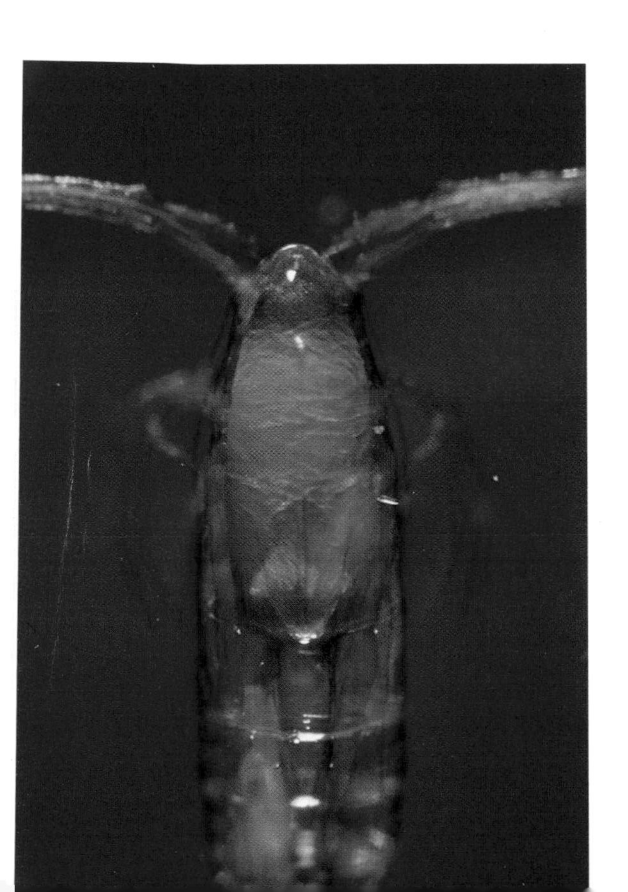

Whale lice

SIMMS LIBRARY
ALBUQUERQUE ACADEMY

The long pectoral fins of a humpback whale make it look like it's flying through the water. Its Latin name, *Megaptera noveangliae*, means "big-winged New Englander."

Do

zooplankton

ever

dive

as

deep

as

a

whale

?

How does a zooplankter avoid the many

Scientists have discovered that at least some zooplankton, in spite of their name, spend more than 80 percent of their time moving under their own power. Many zooplankton are able to move up and down quite well.

Many larger zooplankton, such as krill, take daily journeys into the depths of the sea, often moving hundreds or thousands of feet. Smaller zooplankton make shorter trips of 10 to 50 feet a day. Why do they make such an exhausting journey every day? Perhaps it helps them escape the many predators that crowd the surface waters of the ocean. Or as they move up and down, they may get picked up by an ocean current that brings them to a patch of water with a good supply of food.

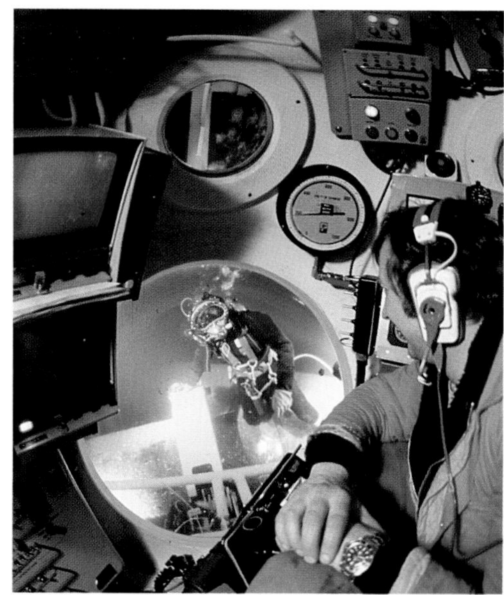

A submersible pilot and a diver greet each other 200 feet below the surface of the North Sea.

small fishes that want to feed on it? *M i g r a t e* **!**

Their up-and-down journeys seem to be controlled by light. The zooplankton move down as the light above increases. Every night, when predators can't see them, zooplankton swim up to the water's surface to feed on phytoplankton and other drifting animals smaller than themselves. As the sun rises, they sink into deeper water, sometimes traveling 1,000 feet or more each way. The change in water pressure is huge and the change in temperature is like going from the equator to Iceland in a single night.

The deeper you dive into the ocean, the greater the pressure you would feel from the weight of the water above you. At a depth of 3,000 feet, the water pressure is enough to squeeze a piece of wood to half its size. Even the most experienced scuba divers can only dive to about 200 feet before the water pressure becomes too much for them. How do tiny sea creatures move up and down through water that would crush us humans?

It is because they are mostly water themselves, which makes the pressure inside their bodies equal to that outside their bodies. Unlike humans, they have no air inside their bodies that can be compressed.

How do zooplankton know how to move up at dusk and deeper with the dawn? Many, such as copepods, have very simple eyes called "eyespots." They can't see images like we can, but they can tell the difference between light and dark. Try this—close your eyes and turn toward a bright light. Cover your eyes with your hands and then remove your hands, keeping your eyes shut. Can you detect a change in the light? That's about what many zooplankton can see.

Some zooplankton can see quite well. They have compound eyes, like insects do, so they see both up and down at the same time!

Krill can move thousands of feet down to deeper water in search of food or to escape being someone else's food.

Can zooplankton really

make a submarine disappear?

Masses of drifting zooplankton can conceal larger objects, like this diver

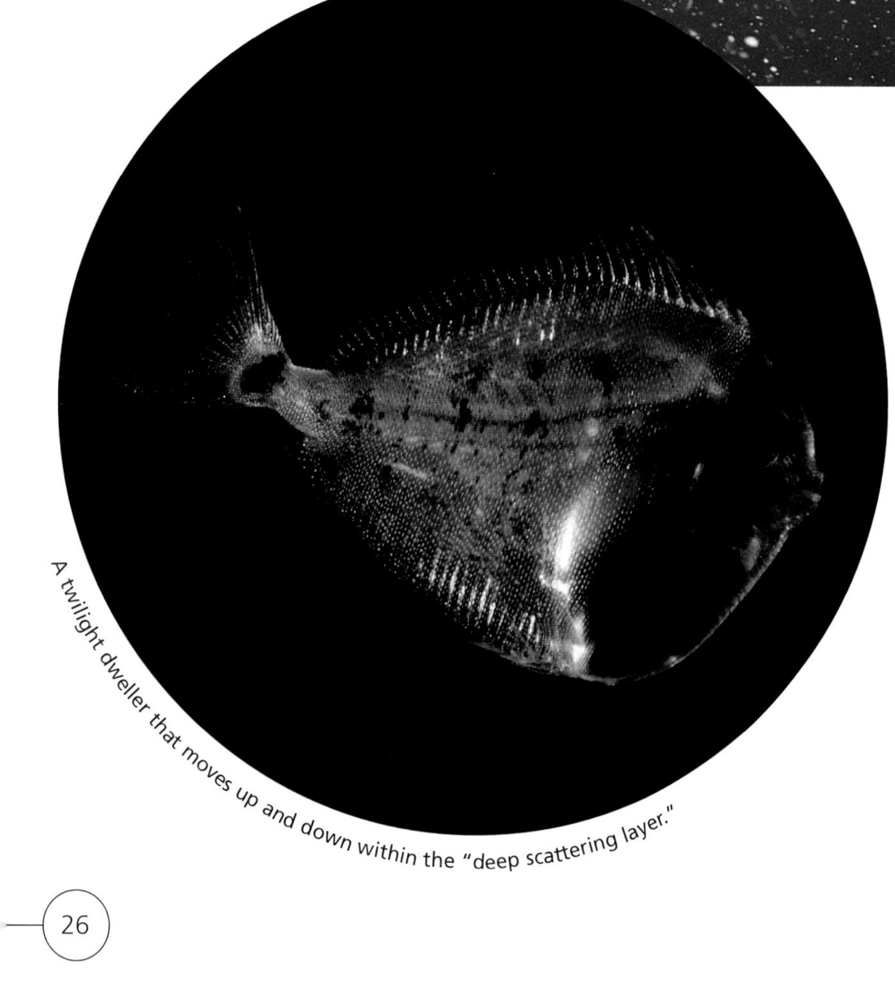

A twilight dweller that moves up and down within the "deep scattering layer."

Sonar was invented to help warships find enemy submarines and to help sailors judge how far the ocean bottom was beneath their ships. Sound waves are sent out underwater, and the sonar records how long it takes an echo to bounce off an object and back to the receiver. Using the same principle as the natural sonar of dolphins, this instrument allows scientists to map the sea floor and fishermen to find schools of fish.

During World War II, Navy ships picked up peculiar readings in the deep Atlantic and Pacific Oceans. Hazy images showed up on their sonar screens that looked like the ocean bottom rising far above the ocean floor. At first scientists thought they might be the tops of underwater mountains. Then they realized they were too indistinct to be solid rock; they were more like drifting underwater clouds.

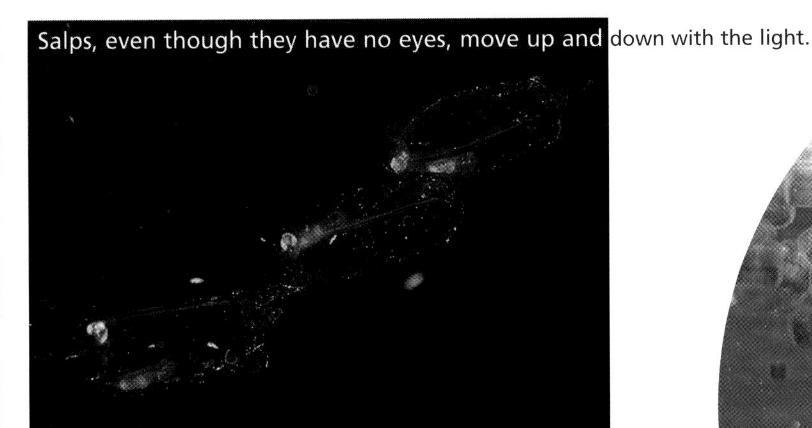
Salps, even though they have no eyes, move up and down with the light.

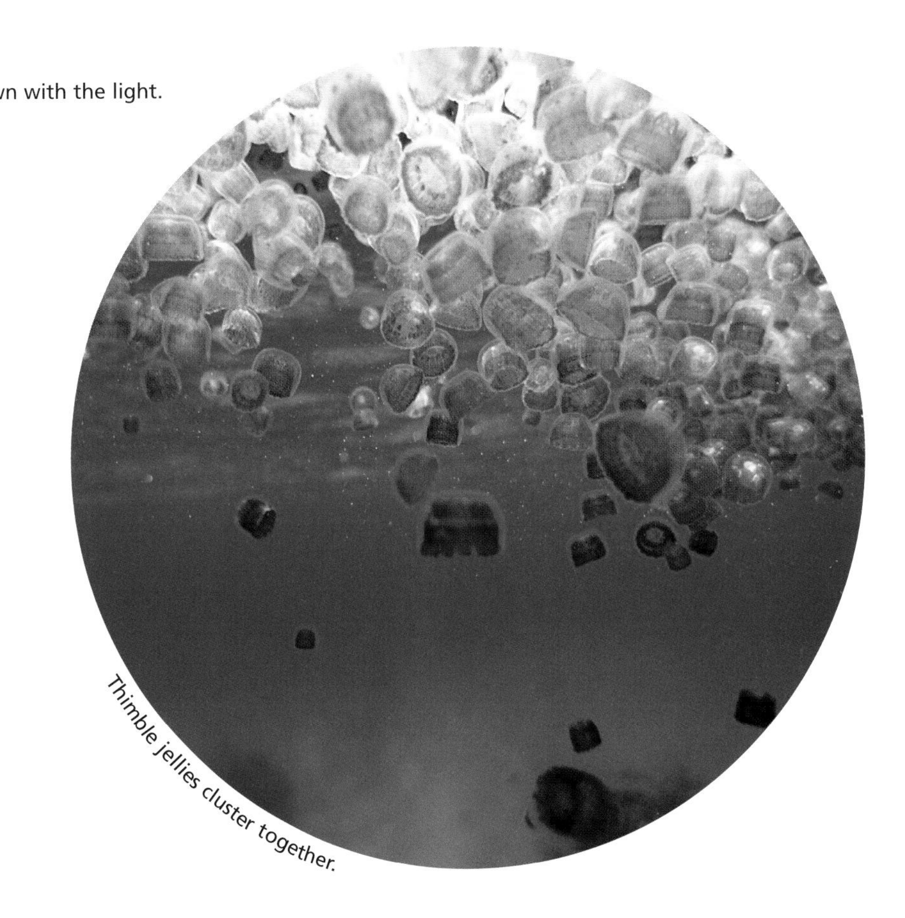
Thimble jellies cluster together.

in their midst.

Navy scientists didn't know what created these "false bottoms," but they realized they could make ideal hiding places for enemy submarines. Because this sound-reflecting layer scattered the sonar signal returning to the ship, this unusual ocean feature was named the "deep scattering layer" or "DSL" for short.

A scientist at Scripps Institution of Oceanography discovered that these layers moved as deep as 700 to 2,400 feet during the day and rose almost to the surface at night. Scientists found that the DSL floated closer to the surface on cloudy days than on sunny days. Whatever was responsible for creating these layers must be alive!

Because of their possible use in submarine warfare, both the Japanese and the Americans kept information about these deep scattering layers secret until after the war. But it took even longer to solve the mystery. Scientists dragged nets through the water but they came up mostly empty. It wasn't until faster, deep-water nets and deep-sea cameras were invented that researchers finally could prove what was making these layers. They found that they consist mainly of small, jellyfish-like zooplankton called siphonophores, krill, and deep sea fish that probably feed on the plankton. These animals occur in such vast numbers that their bodies reflect back the sonar (sound) waves—so zooplankton could actually hide a submarine!

Planktonic shrimp, although they are as clear as glass, can show up on a sonar image when thousands bunch together.

BARE
BARE
BARE

Are there zooplankton you don't want to *bump* into?

This jellyfish is not only colorful, it glows!

You could swim through a sea of zooplankton and never even notice them,

If you were swimming in the South Pacific Ocean and bumped into a sea wasp, you would know it immediately. A collision with this jellyfish is agonizing. And without medical help, you would be dead within a few minutes. The sea wasp (also called a box jelly because of its square shape) is often as big as a baseball, but it can range from the size of a marble to as big as a basketball. Its tentacles stretch out up to 15 feet behind it. Each tentacle has thousands of coiled stinging cells that shoot out like tiny harpoons when they touch something edible.

Jellyfish don't sting humans on purpose. Sea wasps come into shallow water to feed on other zooplankton, shrimp, and small fish, and that's when a swimmer is likely to blunder into the path of their tentacles. At least 65 swimmers off Queensland in northern Australia have been killed by sea wasps in the past 100 years, more than those killed by great white sharks in these waters. Safety nets have been strung across many swimming beaches in Australia to keep out sea wasps, and lifeguards have medicines to treat victims quickly.

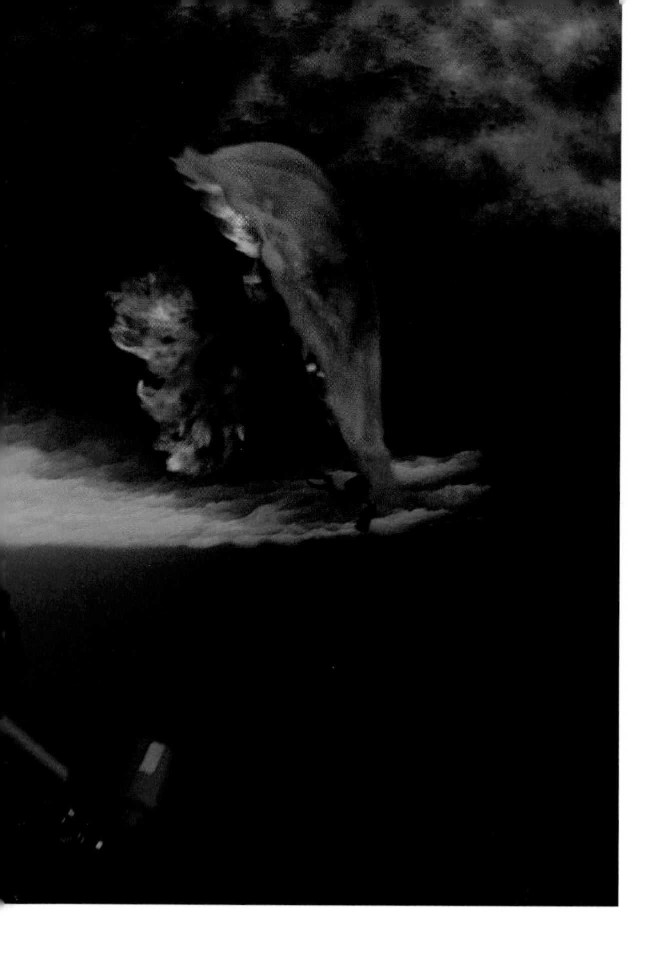

The many tentacles on this Alaskan jellyfish make it an effective predator.

except for a certain few….

Lion's mane jellyfish can grow very large.

Most other dangerous zooplankton also belong to the jellyfish family. There are many different kinds of jellyfish, some as small as a pea and others up to 7 feet wide. The lion's mane jellyfish, for example, has a bell as wide as 3 feet across and its tentacles can be 60 feet long. A jellyfish has no head, no spine, no heart, and is 95 percent water. Its body is shaped like an umbrella that opens and closes to move it gracefully through the water.

Although jellyfish may look delicate, they are powerful killers. The venom from their stinging cells usually paralyzes the prey while long tentacles pull the victim to an opening under its umbrella, which is the closest thing it has to a mouth. In humans, jellyfish venom may cause skin rashes, muscle cramps, and in some cases, death.

Wh

O are the ghosts of the sea?

Pteropods (TERra-pods) are snails without shells. Their name means "wing-footed."

Transparent, unearthly creatures inhabit the

This beautiful golfball-size jelly is a *Benthocodon hyalinus*.

Many of the zooplankton in the deep ocean are "gelatinous" zooplankton, more than 95 percent water, with bodies as soft as Jell-O. These zooplankton are so wispy and delicate that they do not seem to be of this world. They seem far too fragile to be able to withstand the fierce storms of the open sea. Yet here they are, gossamer creatures like the sea butterfly, which "flies" through the water on tiny wings that are actually extensions of its feet. This creature is a mollusk, like snails, clams, and squid. Some mollusk zooplankton have a tiny shell; others have none.

Imagine a transparent barrel a few inches long floating in the sea. That is the salp. Eight bands of muscles running down its side contract to move it through the water and push food and oxygen into its body. Salps often live together in huge colonies of 500 individuals or more that can be many feet long. Yet these salp chains are so delicate that they tear apart if scientists try to capture them in their nets.

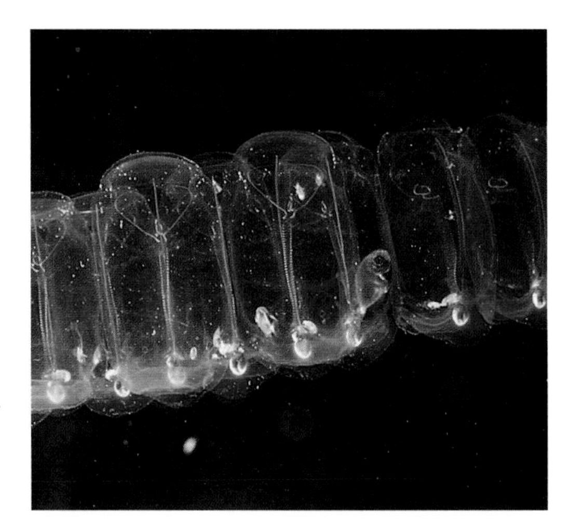

Salps are barrel-shaped gelatinous animals.

pen waters of the ocean far from land.

An arrow worm is transparent except for its two black eyes and whatever it had for lunch inside its stomach. Its mouth is ringed with hooked teeth that can latch onto passing copepods or larval fish. It injects venom into its victim before swallowing it whole. Its prey may end up clearly visible inside its transparent body. (Aren't you glad others can't see what you had for lunch?)

Some zooplankton give off a ghostly glow. Comb jellies look like grapes with rows of colored lights running down their sides. Like fireflies on land, they produce a living light called bioluminescence which doesn't give off heat. It is the same process that is used in the light sticks that light the way of trick-or-treaters on Halloween.

Despite their name, comb jellies are not jellyfish. Instead of stinging cells they have sticky "lasso cells" that can grab onto prey up to ten times bigger than themselves. They have eight rows of comb-plates with tiny hairs arranged like the teeth on a comb. The combs beat in a wave pattern and the hairs row like tiny oars to move them through the water.

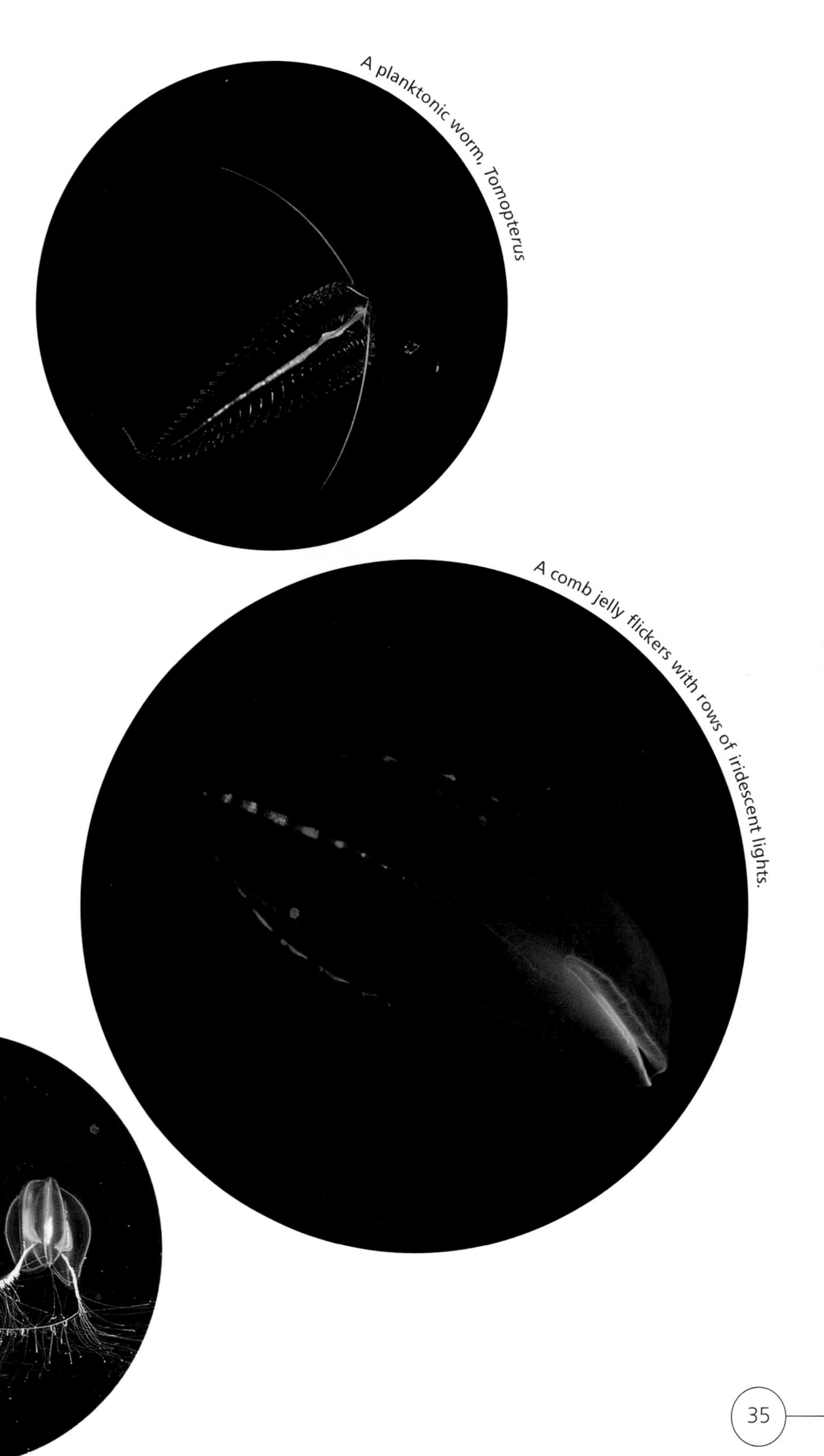

A planktonic worm, *Tomopterus*

A comb jelly flickers with rows of iridescent lights.

A ctenophore (TEEna-for) or comb jelly

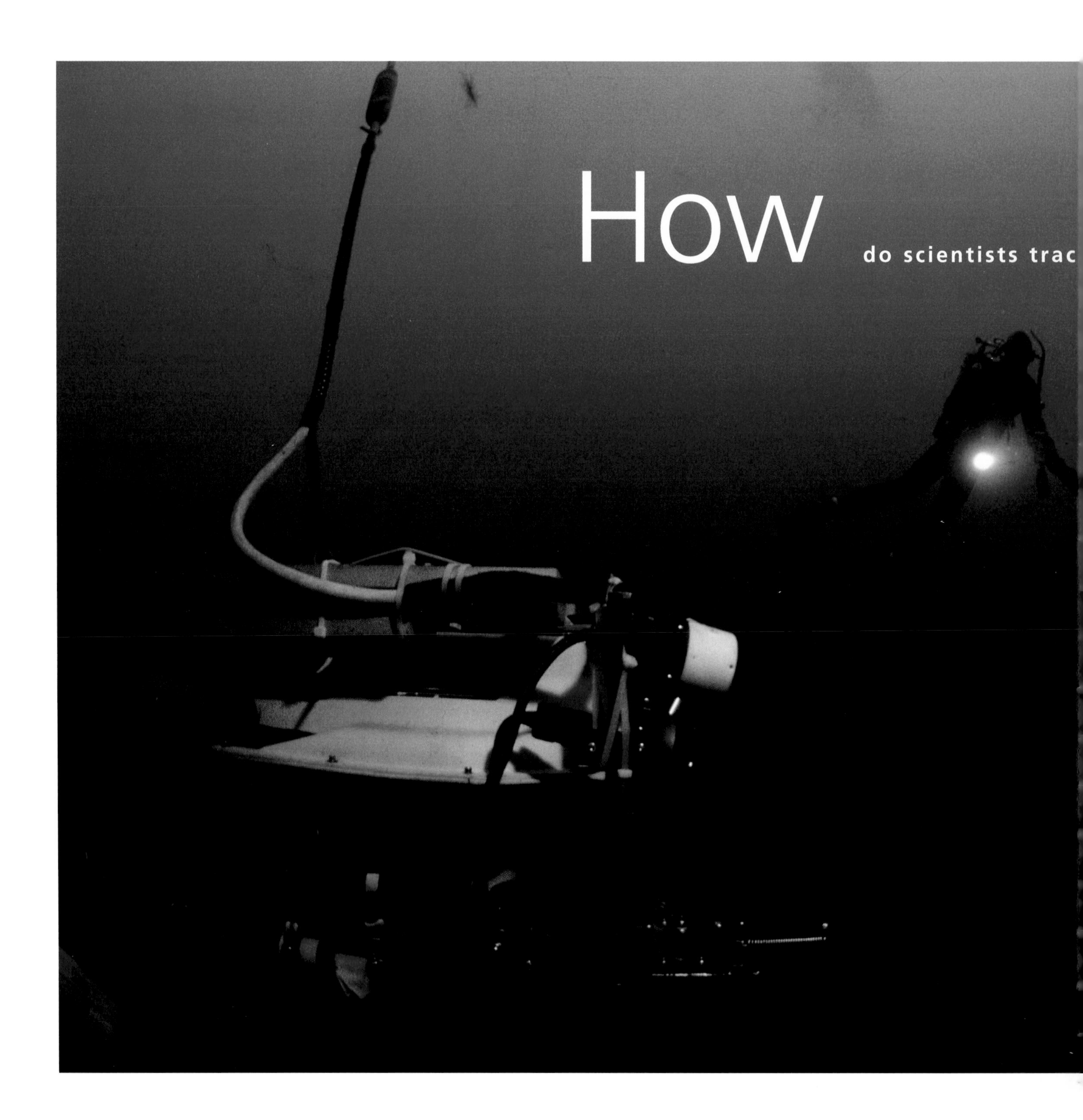

How do scientists trac

the elusive zooplankton?

How can you capture something smaller than a pinhead or more delicate than a butterfly?

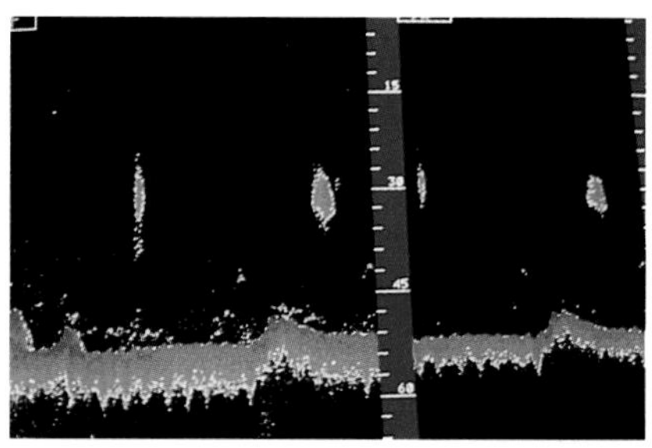

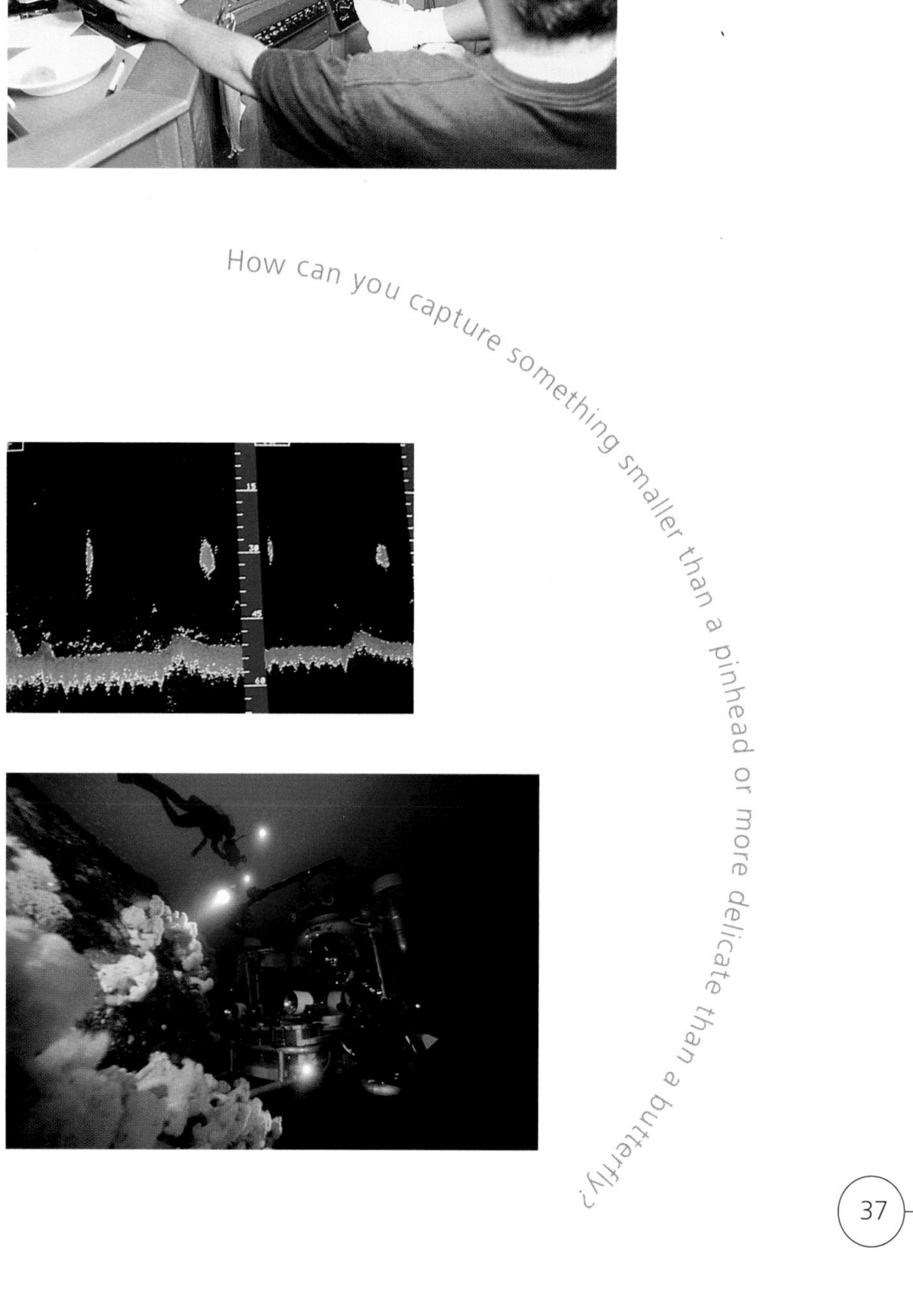

Carefully!

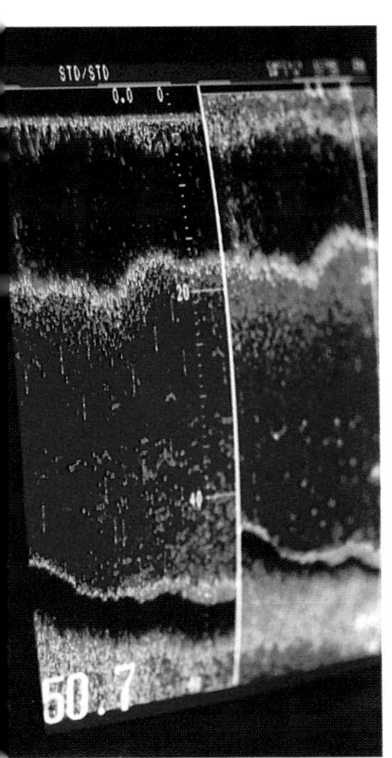

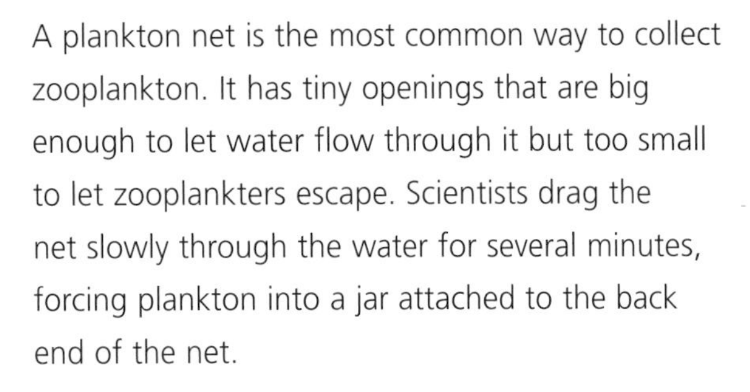

A plankton net is the most common way to collect zooplankton. It has tiny openings that are big enough to let water flow through it but too small to let zooplankters escape. Scientists drag the net slowly through the water for several minutes, forcing plankton into a jar attached to the back end of the net.

Researchers can find out where certain zooplankton live by trapping them in a series of nets hung at different depths. Signals sent from the ship's command post open and close each net automatically. Scientists later study the zooplankton under a microscope and do experiments to learn how much they eat, how many eggs are laid, and how fast they grow.

But plankton nets only catch a small number of the zooplankton in the water and some, especially gelatinous zooplankton, are torn apart or crushed as they are brought out of the water. Some scientists from the Woods Hole Oceanographic Institution have developed the Video Plankton Recorder, an underwater microscope that photographs zooplankton underwater. At the end of every cruise there are many thousands of photographs to review. Fortunately, the scientists have created a computer program that can analyze the photographs, catalog them by species, and select the best shots for the scientists to examine.

Scientists also use sonar to provide details of how zooplankton are distributed in the ocean. All the information collected by nets, sonars, and underwater cameras is used to create a picture of where various zooplankton live in the ocean.

So far, scientists have identified thousands of species of permanent zooplankton, but they know there are many more kinds of zooplankton they haven't even met yet. So the search continues to learn more about these important ingredients of sea soup. Like an old family recipe handed down from generation to generation, this sea soup keeps changing with the addition of each new ingredient.

GLOSSARY

		page
bioluminescence	the production of light by living things, such as bacteria, phytoplankton, and many animals through a chemical reaction	35
bloom	the rapid growth of phytoplankton, often following a flood of nutrients after a heavy rain or a string of sunny days	3
chlorophyll	a chemical in phytoplankton and other plants that captures the light energy and changes it into chemical energy	3
copepod	a very small shrimp-like zooplankter that is a key animal in the ocean food chain	2, 5–6, 11, 15–19
deep scattering layer	a horizontal zone of animals, usually schools of fish or krill, occurring in the depths in many ocean areas. Originally mistaken by some for the ocean bottom when detected by sonar, this layer moves up and down in a daily cycle	26–27
food chain	a linear sequence of who eats whom	3
food web	a complex interrelationship of who eats whom	19
krill	a two-inch-long, shrimp-like animal that is essential food for many Antarctic animals (and elsewhere)	2, 11, 13–15, 22–23, 27, 38
larval	the early stages of many animals between hatching and adulthood	10, 11
migrate	to move from place to place	22–23
mollusk	a group of soft-bodied animals, many of which have shells, such as clams, snails, octopuses, squid, and sea butterflies	32–34
nutrients	simple chemicals required for plants and animals to live and grow	3
photosynthesis	process in which light energy from the sun is converted into chemical energy by plants using water and carbon dioxide, resulting in the production of oxygen, sugars, and starches	3
phytoplankton	plant plankton, which harness the energy of the sun to make their own food	3, 6
plankter	a single, floating plant or animal	3
plankton	floating plants or animals that drift with the current or have very limited swimming ability	3
predator	an animal that eats other animals	15, 23
prey	an organism that is eaten	15
salinity	the amount of salts dissolved in water	
sonar	a device for detecting objects underwater by the reflection of sound waves	2–3, 5, 7, 11–12, 22–23, 26–27, 37–38
zooplankton	animal plankton that feed on phytoplankton or other zooplankton; can be permanent or temporary zooplankton	

BIBLIOGRAPHY

Baldwin, Robert. 1998. *This Is the Sea That Feeds Us.* Nevada City, CA: Dawn Publications

Cerullo, Mary. 1999. *Sea Soup: Phytoplankton.* Gardiner, Maine: Tilbury House

Cerullo, Mary. 2000. *Ocean Detectives: Solving the Mysteries of the Sea.* Austin, Texas: Raintree Steck-Vaughn Publishers

Kaufman, Les and staff, New England Aquarium. 1991. *Alligators to Zooplankton: A Dictionary of Water Babies.* New York: Franklin Watts

Kovacs, Deborah and Kate Madin. 1996. *Beneath Blue Waters: Meetings with Remarkable Sea Creatures.* New York: Viking

Sibbald, Jean. 1986. *Sea Babies: New Life in the Ocean*. Minneapolis, MN: Dillon

Tilbury House, Publishers
2 Mechanic Street • Gardiner, Maine 04345
800-582-1899 • http://www.tilburyhouse.com

Gulf of Maine Aquarium
P.O. Box 7549 • Portland, Maine 04112
207-772-2321

Text copyright © 2001 Mary M. Cerullo.
Photographs © 2001 Bill Curtsinger.

First printing: February 2001.
Hardcover 10 9 8 7 6 5 4 3 2 1

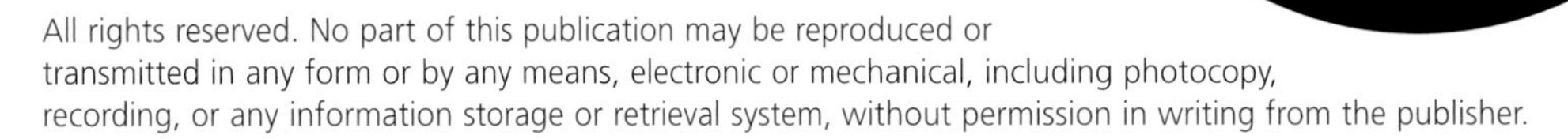

All rights reserved. No part of this publication may be reproduced or transmitted in any form or by any means, electronic or mechanical, including photocopy, recording, or any information storage or retrieval system, without permission in writing from the publisher.

- This book is dedicated to
  My sister Nancy, who nurtures blooms on land —M. M. C.
  Don Perkins, the great crusader, encourager, and motivator —B. C.

- Our thanks to Dr. Ray Gerber, St. Joseph's College; Dr. Bernie McAlice, Tim Miller, and Kevin and Pam Eckelbarger, University of Maine Darling Marine Center; Dr. Charles Gregory, Southern Maine Technical College; Joe Payne, Peter Milholland, and Mike Doan, Friends of Casco Bay; Michale Hallacy, Carl Zeiss, Inc; Joe Marcantonio and the F/V STARLIGHT; Lew Incze, Bigelow Laboratory for Ocean Sciences; and Doug Volmer.

Library of Congress Cataloging-in-Publication Data
Cerullo, Mary M.
Sea soup : zooplankton / Mary M. Cerullo ; photography by Bill Curtsinger.
p. cm Includes bibliographical references (p.).
ISBN 0-88448-219-7 (alk. paper)
1. Marine zooplankton--Juvenile literature. [1. Plankton.] I. Curtsinger, Bill, 1946– ill. II. Title. III. Title.
QL123.C42 2000 00-046721
592.1776—dc21 CIP

Design and layout: Geraldine Millham, Westport, MA • Editing and production: Jennifer Elliott, Barbara Diamond
Color scans and film: Integrated Composition Systems, Spokane, WA • Printing and binding: Worzalla Publishing, Stevens Point, WI

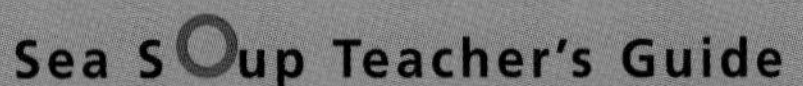

Sea SOup Teacher's Guide
Discovering the Watery World of Phytoplankton and Zooplankton
by Betsy T. Stevens, illustrated by Rosemary Giebfried

Paperback, $9.95 — 96 pages, illustrations
ISBN 0-88448-209-X — Children/Science Grades 3–7

The interesting and fun activities in Betsy Stevens' *Sea Soup Teacher's Guide* meet the challenge of relating tiny, microscopic organisms to the lives of children. Discover and explore answers to some strange questions. What is the recipe for Sea Soup? Are those tiny critters plants or animals, or maybe something else? Why do they look more like creatures from outer space than the organisms we know on land? What do giant clams, corals, whales, penguins, and humans have in common? How does the Sea Soup grow? What if it stops growing?

The inquiry-based activities range from designing and making a phytoplankter and collecting phytoplankton to designing an experiment for exploring what factors influence the growth of phytoplankton and zooplankton. The emphasis is on science, but where appropriate math, geography, language arts, and art are included. Each unit includes background information, objectives, a statement of how it addresses the National Science Education Standards, materials, procedures, references, and suggested websites.

SIMMS LIBRARY
Albuquerque Academy
6400 Wyoming Blvd. N.E.
Albuquerque, NM 87109